WILD HOGS

Stories From The Field
Vol. 1

Author
Richard M. Schamber

FORWORD

Wild hog hunting today is growing state by state day by day. Other than in the deep south many hunters are very new to this hunting sport. For over thirty years I have been lucky enough to have hunted these wily and most often elusive critters. Because I grew up hunting in the state of Florida wild hogs are an everyday part of the hunting experience. I'll be the first to tell you that hunting an old seasoned boar is just as hard as any trophy whitetail deer I have ever been up against. My hope by telling these stories from my years of hunting wild hogs I can enlighten and entertain hunters to the great sport I have enjoyed for many years.

Richard M. Schamber

DECICATED TO

This books as all my books are dedicated to my father he was a great inspiration in my life as well as my out doors experiences. He was a great hunter and lover of all things wild. In his lifetime he was able to watch as wildlife of all kinds and habitat was restored and increased to numbers far greater then expected. This he said warmed his heart.

U.S.A.F. Retired
Col. Stanley Q. Schamber
1912 to 2003

Table Of Contents

BIG RED

Some years back I had gotten on a lease in south Florida that was an awesome place for wild hogs. I had hunted this place several times on two day hunts always with success. When the chance came for us to lease it for the archery season we had to jump on it. The weekend before the archery season we were allowed to come in and set up camps and our stands. If we wished to hunt for a small fee we could. Since I love hunting hogs this I did. It's always the labor day weekend we do this because the archery season in south Florida starts the weekend after. This being a long weekend makes it easy to set up camp and stands leaving plenty of time to hunt.

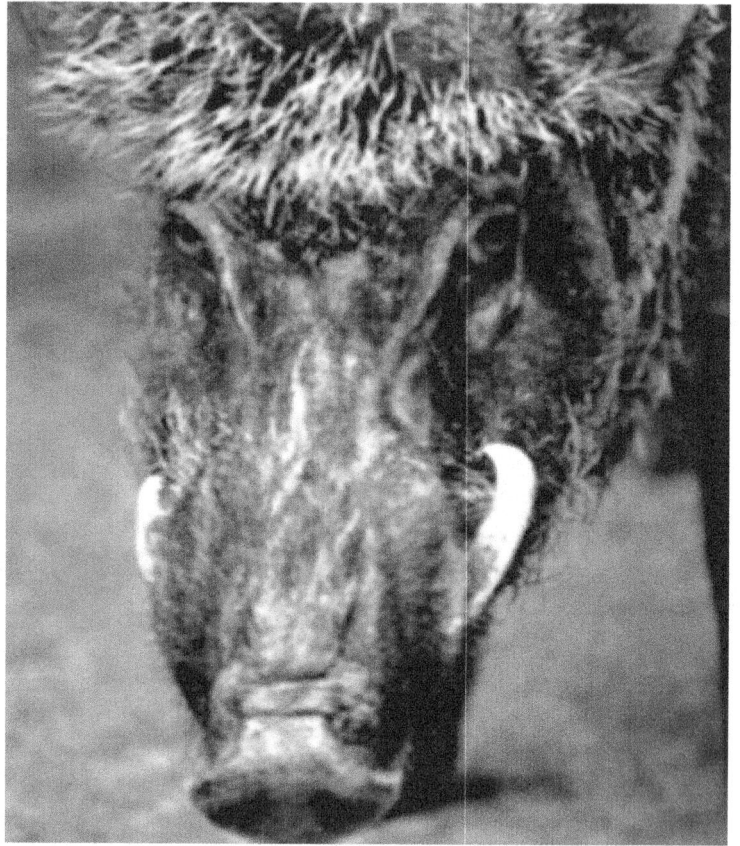

Just two months before I had been on two two day

hunts that were archery only and just for hogs. This is when I first saw Big Red. He showed himself on both hunts but never in a situation where a shot at him could be taken. I corned up the roads where he had been but this didn't do any good. For some reason he moved in the middle of the day like an old buck deer. Never at times when most hogs do.

Red was a big hog long, lean and mean with the big teeth to match. All this matched his disposition and more. There was no good way to tell how much he really weighed but I was figuring well over two hundred pounds. With any luck at all maybe soon I would find out.

One day around noon on the second hunt while coming back from scouting a stand location for no good reason at all he shot out of the brush right at my truck. First you think hes just running spooked and will pass right by. This was not the case. He ran right in front of the truck stopped and stood me off. Yes that's right he stood off my truck as if to say how dare you come into my territory. He looked mean and mad I had interrupted his day. I was very happy to be in my truck and not on the ground with him. I had heard of big hogs going after atv's, cattle and horses but come on now a TRUCK ! There was no hurry and felt my truck could win the fight if need be so I just sat and waited him out. I'm sure it seemed longer than it was before he finally moved away still I'm guessing it was a good six to eight minutes before he did. Oh and yes there was a witness!

That was on the second June hog hunt and I never saw him again that weekend. Driving home after the hunt I kept thinking about him and how bad I wanted another chance to hunt him. We had the land leased for archery so it would be my intention to find him and take him with my bow. No small task that was sure but was something I had to try and do. The next two months were full of dreams of how this could be accomplish.

Our labor day weekend started on Friday at noon and ran through Monday noon. So plenty of time to do all we needed to. I arrived shortly before noon that Friday got my lease member card payed for my three days of hunting and set out for camp. Pumped up to get started was an understatement.

The temperatures out were still hot but this didn't kill my enthusiasm to get out and hunt that evening. My plan was to first find as much sign as possible then spread out some corn. The goal was hunting hogs not just Big Red. There were five weeks to do that. We were allowed two hogs on this weekend so I was going to do all I could to make that happen.

My spot was in the south eastern most corner of the

property. It was a long drive from camp but well worth the distance. I had hunted hogs in this area many times and knew my chances were very good. Once all my gear

was together I set out. For some reason I headed dead straight to the spot of the first encounter with Big Red. There was an oak hammock that kind of surrounded a grass pond that often had hogs feeding and travailing through it. If there was the sign there I hoped to see I would put out some corn and look for another spot to do the same. On a hog hunt like this I try to cover as many areas as can and for the most part still hunt. Moving around the hammock there was plenty of sign so I headed back to the truck for the corn. After stringing some out I set out for another favored spot along a fence line. In the past I had strung corn along this fence for up to half a mile with great results.

There is a massive slew on one side of the fence the hogs love to lay up in during the day. When it gets to feeding time in the evening they cross the fence in several places. This makes it perfect for still hunting. Starting where I normally did I strung corn a long ways covering many hog crossings. When done it was time to park my truck and off I went. It was early so no need to get in a hurry. Mostly I worked through some hammock country to waist time but to look for tree stand locations as well. Once the air cooled down some I headed for the fence. Something was bound to happen.

This fence line was long and straight and you could see down most of it. All I had to do was be at one end to watch and wait. I could hear hogs squealing in the hammock I had first put out the corn in. It wasn't far so if soon I didn't see anything on the fence I would head to the other hammock. It sounded like there might also be some fighting going on maybe because the sows were coming in heat. There was plenty of time so I made no move until I was sure of what to do. I really didn't want to move away from the fence. I had so much luck there it was like a magnet. Keeping track of time and it getting late it was about time to head to the other spot before it got to late to get there and make a stalk. Just as I stood up down from me four big hogs walked out on the fence line. Right away they found the corn and started feeding. It looked like about two hundred yards to them so I had to go. There was plenty of time but couldn't waist any of it. The stalk was on.

I felt the best way to them was to slip up into the hammock and use it to cover my movements. This could get me to within twenty yards of them if I played my cards right. Slipping along was easy and soon could see them through the trees. They had moved in my direction some and were still heading along the fence. Them being totally distracted by the corn would give me what I needed to go the extra few yards to be in range. Moving to my right I got behind an oak and waited. Still heading my way feeding there was no need to get closer. The shot I would have if this worked out would only be ten yards and perfect broadside. All four of the hogs were boars and in the hundred and fifty pound range. Perfect for the table just what I was looking for. One was about to be mine.

Just as I had hoped for they moved right along the fence to me one gave me the right shot and I took it. The arrow made a pass through in the right spot and my hog only ran about forty yards and expired. He never left the fence just ran along it and fell over. The hunt couldn't have worked out any better not only did I get my hog but it could be driven to.

At camp that night I was thinking about how one would hunt the hammock where I had heard all the excitement. Figuring I had two days to make something happen there so why not go for it. This would also put me right in the middle of Big Reds territory. What I had heard that afternoon was surly breading activity if this was the case he might just be close.

The next morning I headed out extra early to allow me time to string out some more corn. I wanted to cover all the bases in case my hammock didn't work out. Knowing I couldn't go into the hammock just to put out corn and then still hunt it I put out the corn along the fence line where it had been before. My idea was to catch hogs heading back to bed and maybe stopping them long enough to get a crack at them. I would know right away if the hogs were back in the hammock and if not off to the fence it would be. Once done with the corn I parked and slowly headed to the hammock.

The sun was up just enough so I could see and to know what was ahead of me. I hadn't seen a single hog out anywhere so far not even on the drive in. It was a little warm but not so it should keep them from moving. Surely I would come across some on the corn from the night before.

Arriving at the edge of my little hammock I could see right away the corn was gone from where I had started putting it out the afternoon before. That didn't surprise me to much it had been fifteen hours. With a hundred yards to stalk through to the other end all one could hope for was maybe the hogs were somewhere in the middle or the other end. Watching ahead slowly I

moved along. Still not a single piece of corn on the ground they had really cleaned it up. Just about when it looked like there was no chance to see them I caught sight of something move. That movement came from a small hog searching for more corn. I got honed in on it and watched. Within a few seconds a second showed. Soon a third appeared all the same size and very small. Seeing that they were no more than forty or fifty pounds I had no interest in taking one. After watching them for a few minutes it was clear it was just the three. Letting them move off I slowly went the other way and slipped back to the fence line. Time for plan B.

Along the way I walked up on two deer and yes did proceeded to spook them. Yes you guessed it they went right for the fence. Not exactly what I wanted but never the less I had a mission. Thinking if any hogs on the fence had noticed the deer running past it might be best to wait a few minutes before stepping out. I gave it a little time before I did.

The chosen spot to step out in was a patch of brush with a couple of palm trees that grew right on the fence line itself. This cover would be helpful in concealing me. I glassed down the fence and sure enough there were several hogs feeding on my corn line. Right away I was off time was short it was warming up fast and there was quite a distance to go. Just like the evening before I hit the long hammock that paralleled the fence. This would be used to cover my movements and make my stalk. The wind was up slightly and swirling never good on a ground stalk. I worked my way close and could barely see them just about a hundred yards ahead. I moved closer very slowly and quietly. Now within fifty yards I could see them well. The one thing that had changed was the wind. I could feel it on the back of my neck heading right at the pigs. Wild hogs have a great sense of smell making it hard to get close in a bad wind. I kept on using a large oak to block their vision of me. It was working until I was about thirty yards and like a shot from a gun one looked up woofed and all ran like hell. This happens and one should never be discouraged there will be more chances just give it time.

The weather was warming up fast I knew the hogs probably wouldn't be back out to feed until it cooled of that evening. Time to continue with what I would normally do and that was to go back start all over and re-string all my corn lines. At worst maybe a deer or some turkeys might feed on it but not enough to make a difference. I knew it would be a long afternoon as I still needed to hang stands in locations for the coming archery season. So back to camp it was to regroup and get it done.

When I arrived in camp most of the twelve members of the club were in and not but about four had hunted hogs that morning. Two had taken nice size pigs for eating and had seen plenty more. For the most part the rest were there for the deer hunting to come the following weekend. We had a large map for everyone to mark their stand locations on. This is always helpful to inform others of these locations preventing all sorts of problems. No one was anywhere close to the spots I was hoping to place stands in so it was sure I was in good shape to do so. All these spots would be good for deer, hogs and turkey so the work was about to begin. So with the spots marked I wanted time for a lunch take an hour nap and it was off again.

With each stand I placed a feeder one was a bump feeder the others were automatic. Putting them in early would also give me alternates for the remainder of this hog hunt if the hogs were to find them fast. This as we all know is the work part of hunting but if done well it pays off in the long run. Take your time do your scouting read the sign and sometimes go on your best instincts. We all have them and quite often I do that with success.

The early afternoons work had been longer and much hotter then I had expected. Once it was all finished considering the long drive back to camp and time left in the day to hunt I opted to stay out. I had extra hunting cloths in the truck so I just relaxed for a bit changed cloths and hit the trail. I didn't get far and the regular afternoons storms rolled in so back to the truck it was. It rained hard for about forty five minutes. This was a welcome relief from the heat and with the amount of rain that fell the temps should stay down making the chances of seeing hogs before dark better. Ready to go off it was once again poised for action.

First I went to walk the corn I had put out along

the main road into my area. I had strung that line right down the middle of it for almost a half of a mile. With many bends and twists in this road its one of my favorites to stalk down. You could stalk along with easy walking and see ahead well. Rounding a bend I did come across two deer feeding on the corn this was a good sign. If it was cool enough for them to feed the hogs shouldn't be far behind. I sat down right there and just watched them. I had a stand close to there and was interested to see if a buck were to come out to feed as well. Our lease only allowed us to take bucks no does at all so knowing all I could about their activity was important. The does fed for a good twenty minutes and left. Once they were out of the picture I moved off to that end of the corn and started back the other way.

While moving back down the line my gut was telling me to go check the small hammock where I was sure Big Red liked to hang out. I wasn't sure if it was the right move but something was making me go. This time I would come in from a different angle using the wind. My plan was to move into the small gap separating the front half from the back half of the hammock. There was a small dry grass pond I would have to get past before reaching that gap. The wax myrtles along the edge would give me cover and ease of movement. As I was passing some oaks and palms just before this grass pond

I heard some sounds coming from my hammock. Sure enough it was hogs and it sounded like fighting during competition for breading rights. This is just what I had heard the night before. It was sure with this going on in Big Reds area he would have to be in on it.

Moving slowly along the pond I went. You could hear the action well and they seemed to be moving to the gap I was also heading for. Half way around the pond and about a hundred yards from it the sounds they were making said they were almost to the gap. The fighting was intense very loud and non stop. This had to be several big boars haven at it. The fighting was a good distraction they would be so intent on the fight maybe I could get close with no effort. At seventy five yards away I used a lone palm tree to cover me. From this point on I would be in the open until about twenty yards from the center of the gap. I just started around the palm to make my next move and there

he was in all his splendor Big Red. Well let me tell you right now my nerves went to hell and I mean fast. When he had stood me off in the truck I knew he was big but until your on the ground with them you really never know just how big. In no way was I wanting to make the wrong move here and now. Wasn't feeling like being hog food on this day. He looked right at me he knew I was there. It was almost like he had come into the opening just to have a look at me. It was an eery feeling I knew something in his sixth sense told him I had arrived. I stayed still with hopes he would disregard me as a threat and move past the gap. In just a couple minutes he did. As he moved off he looked at me as if to say to me you don't worry me a bit.

Now I had to catch my breath this was intense he was close. With him out of sight I moved in ready to draw my bow if needed at any moment. I'm a good snap shooter but there is always some question if you can rise to meet the task in a split second. My last cover was just ahead and again a medium sized palm tree. Would this be enough or would I have to go in after him? Would I be able to with all the other hogs so close? Hiding behind the palm I could hear lots going on to my right more fighting and loud squealing. Big Red had moved off to my left would this fighting bring him back through the gap, I wasn't sure. I could hear walking from the right sounding like another large boar. It was a big black boar and maybe close to two hundred pounds. I'm thinking maybe I should take him and wait for another chance at Red when the terms were more in my favor. The big black moved right in stopping at fifteen yards broadside a perfect shot. Just then came more loud squeals off to the right he looked and moved right back that way. What happened next was as if scripted Big Red walked right into my line of sight. He stopped twenty yards slight quartering angel away giving me the right angel to slip an arrow in behind the shield. I drew back he looked right at me as I did. He didn't seem to be alarmed and was more interested in the fighting going on. The second he turned his head to look that way I took my shot. The arrow was true the flight dead on I had made the right choice. The arrow sunk all the way in passing right through all the vitals and stopping in the opposite shoulder just as it should be. What I didn't expect was for him to just stand there and look back to see what had just happened. For a split second he just stood there and then slowly moved away. I thought my nerves were about as shot as could be before the shot but after no way could I have moved or reacted to anything. So I was more then happy he just went away from me as if nothing had happened.

I had plenty of time to wait and if needed I would

wait till after dark, no need to hurry this up now. I had him just needed to be smart. Moving into cover I gave it about thirty minutes before even walking in to the impact site. As I expected there was no sign at all not even a drop of blood. I had been down this road before so tracking was the next thing to do and very slowly to look for any other sign I could find. A large hog such as Red would have to leave hoof prints in the moist soil. So slow and go it would be. The main road was at the end of the hammock Red was heading through. It seemed it may be the first best chance to find any indication of what he had done. I had his track and was right on it all the way to the road and still was looking for blood. Right before coming out on the road there it was the first blood and it looked like the entrance had not plugged as I had first feared. Stepping up on the road I found where it looked like he had stopped to sort of get his bearings. He headed into the timber on the other side of the road giving me easy tracking ground to follow. I took a few minutes to glass ahead and see if I might see him down in sight. Also this was to give him just a little more time to expire. This was a really big boar that hadn't gotten in a hurry so the extra time would be of benefit. With still plenty of time in daylight I wasn't losing this hog no way no how.

On the road I hadn't been able to see him but now

had blood and track to follow so on I went. It would be another hundred yards to the fence line and the beginning of the dense cover he was surly heading for. My hope now was to find him before he got that far. If not the real and dangerous work I dreaded would be next. The trail was hot, tracks well defined and good blood. He had to be down the question was where. I moved along ever so slowly watching and learning from the sign I had to work with. It told me this hog was done but he sure had gone a long way. Now another fifty yards down the trail from the road I found another place he had stopped this time he changed direction. He went straight right, this was good for me it would take him into more open country. Often times when an animal hit hard and dieing changes direction such as this it shows confusion. I feel from blood lose to the brain again good for the hunter to know on a long blood trail.

As I moved into the more open ground my hopes heightened the blood trail was now constant and heavy. Watching ahead intently I knew not to far ahead this big boy would be laying there. In just a few more steps it happened I saw him, there he was no more then fifty yards in front of me down and dead. What a sight this was and what a relief as well. These big hogs can be as dangerous as any critter on earth when hurt so approaching with caution is a must. Once to him there was no question he was done and mine the likes of which I had never been this close to in my life. What a day, what a hog, what a hunt. The feeling of great self satisfaction is sometimes overwhelming when a plan comes together. This was true in this case. How lucky was I to have taken such a fine trophy on a spot and stalk hunt. I almost didn't know what to do next all I could do is just stand there almost numb from the whole experience.

It took a good half hour for reality to set in and the realization I still had to load this bad boy up some how and get back to camp, the evening light was fading fast.

This hog had to out weigh me by double my weight and my skinny butt was haven one hell of a time getting him loaded. After a great deal of effort and time I did manage to do it and off to camp I went. I was so very proud to be a hunter following my dads way of life and the lifestyle of so many before me.

I have taken many hogs since that time and not one has come very close overall stature. The scale in camp only went to three hundred pounds and it bottomed out. When I extracted the teeth I had a total of eleven inches per side by the SCI scoring system. And let me tell you one ugly looking hog to mount. Most of all an experience of a lifetime and a story to share with all. Isn't that what its all about? So pursue your dreams and go after your BIG RED.

THE CONQUISTADOR

For many years I had heard about the rare but still remaining survivors of the old Spanish hogs that run the woods in parts of Florida. These hogs are often called Conquistadors. The two main ways to tell them apart from all other hogs is the very long narrow snout plus they don't have split hooves. Another characteristic is that mature boars never really get much larger than a hundred and seventy five pounds in the wild. I had seen the tracks of this rare hog for the first time on our south Florida archery lease and would often wounder if I might ever take one.

For going on six years I had in some way hunted the same lease. Some years it was on two day hug hunts and some years we leased it for the archery and muzzle loader seasons. Our hunting club consisted of the same members every year with all of us having our favorite hunting spots. Mine was the south eastern most corner of the lease that members now call Rick's corner. Most all of the lease was flat full of game and my corner was not really better. Mostly it was just always were I felt was most comfortable being and hunting. I guess as they say " it fit my eye" and I loved being there.

In my little corner I had three tree stands set up plus as an extra, I would from time to time spread out whole corn along the roads or the fence line. One of my favorite ways to hunt the corner was to spot and stalk with my bow. One of the very best ways to do this for hogs is to run corn for long distances in a thin string and later go back to still hunt it. So far this season I had taken two real nice hogs hunting this way.

This was the formal archery season so we were allowed to take 3 Deer, 2 Turkey and 12 Wild Hogs. Most of my stands were set up for both deer and hogs with deer being the primary interest. The deer in the area would come to the feeder/stand sites much better in the afternoon then mornings. This left most mornings open for hunting along strings of corn. I guess looking back on it that was the best of both worlds.

My corner was just that with one fence line running east and west the other north and south. This year the north south fence was working out better for corn lines.

It was along that fence I had taken the first two hogs of the year. What I liked to do was get there well before sun rise run a line of corn about a mile along that fence. I could then go park the truck well out of the way and slowly slip along and watch for the hogs to come out find the corn and feed. Some days this worked well and some not so much. What made this work so well was one a good number of hogs and two was the hammocks that lined the open area off the fence. Along most of that stretch of the fence the hammock was never more then thirty yards away from the fence itself. This made for ease of slipping along hidden and still have feeding hogs in range without having to be out stalking in the open.

One Saturday morning I did my normal thing and strung out my corn. Went and parked my truck then back to the fence. It almost never takes to long and a hog or group of hogs will be traveling and find that corn. This morning nothing was coming out anywhere. There were acorns falling so that was a big part of it. Still even when they were falling the hogs would at least feed some while on their way back to nest for the mid day. This day there was none of it at all. After about an hour and half of it I got antsy and had to move so move along the corn line it was. Not more than a hundred yards down the line I saw the figure of a single hog moving in the direction of the fence. This hog was coming from the opposite direction I was expecting but non the less coming my way. It looked to be good sized and was one of my favorite coloration's. He was a yellowish base color with brown and black spots. An awesome looking hog for sure. The stalk was on.

He was coming across a field that was grown up almost to the top of his back. Most of the ground he had to move through to get to the fence the tall grass would make it hard for him to see me move. This was looking like it might be a simple task to accomplish. I was betting on the fact he would come across the corn and go to feeding. I would then make a final move and get my shot. Well sounded good at first but this pig had other ideas. When got right to the fence he went into super alert mode. I could see he was very suspect of everything even crossing the fence. He would move only a little at a time then just watch. Once he got to the fence he stood there for I know ten minutes watching all around him. No way could he see me and there was nothing in sight I could see that would keep him on edge like this. I just settled in behind a large pine tree and waited for him to make up his mind and commit to crossing. This might be taking awhile.

He did take his own sweet time to cross and when he finally did he right away went to work on the corn.

I felt it necessary to let him feed for a little bit before making a move. This would allow him time to calm down become engrossed in the free food and forget about his surroundings. You know a pig loves a free meal. He had a full line of corn in both directions so there was no hurry. What I really hoped was he turn my way and feed past me then I would only have to be ready and make the shot. Well he had other things on his mind it seemed because he did just the opposite and started feeding away from me. If he were to go more than fifty yards It would be time to make a move. At that point he would be going around a bend in the fence and out of sight. This I was not about to let happen.

Deeper into the hammock I went with the plan being to get ahead of him to set up for a shoot. The oak and palm hammock was slightly thinner there so I would also be able to keep an eye on the hog as I moved. At the bend there was a thin island of trees where the road split. That island was about six to eight yards thick and twenty yards long. I had corned both sides of it and if I could get to the game trail that lead out of the hammock to it I had a way there. Right on the edge of the split the brush got thicker so cover was not a problem. This was the plan now to make it happen.

Slowly I moved along stopping to keep track of him as I went. He was easy to see because of his coloration this helped as I moved. He wasn't moving very far very fast I was gaining ground quickly. Just as I got to the game trail that led to the split in the road he reversed on me and started feeding back to where he had come from. Ok now back up some and start all over it was. He moved the other way quicker then he had first when he first started his original direction. I had to move fast to get ahead. The problem now was if he moved to far back that way I would be out of range and had an open gap to cross to make up the distance. If he got that far I would have to wait until I could cross that twenty yard gap before my next move. This was becoming a cat and mouse game real fast. That darn hog did move to far I was out of position and options I had to wait him out. Because he had come from the east I thought he was sure to move to the west into the hammock once he finished feeding. It may be my last chance to get a crack at him. Well sure enough he did exactly the opposite of

that and went back the way he had come from. My morning was over it had been fun and maybe another day I might see him again.

Once back in camp I was thinking that I had another stand I might just have to put it up on the bend on the fence and hunt in the morning there. Maybe this would get me another shot at him. Right after lunch I went and did just that and put up a stand in that bend. I knew this was also a spot that the deer would use to cross so it could have a duel purpose. I could also see a long ways up and down the fence to help in learning if there were other spots I would need to look at. So interested in this new spot I was that on this evening I would be sitting right there to hunt. Before leaving the new stand site I put out a new line of corn to have it ready for the evening sit.

At this point I had not sat on one of the fence lines in the evening. In the past I had always had luck along them still hunting corn lines. It would be interesting to see what might be happening there now. These fence lines are crossed all day long. Most of the time it's deer during mid day doing so and that was fine I wanted to be there early just in case.

Around 2pm I was parking my truck and on my way to the stand. It would be a long sit with it still being day light savings time. I didn't care this was a new endeavor and a new fact finding mission. The way I set up the stand was to use the little island in the split of the road as cover from what might come from across the open marsh and field in front. Behind me I had plenty of cover to keep me concealed and just enough visibility to see anything coming from in back of me. The grass was still very green so being able to see would be all important hearing things coming would be non existent. I think I really expected to see deer from there more then the hogs right off. Our deer on this lease were not hunted hard with the pressure we put on them almost nil. They should move all day as normal. I know at lease up to this point this year not a single person had hunted in my area but me. Where I had been hunting was not close to this new stand. I fully expected to see movement all afternoon. Not long after I was settled in I did see the first deer. It was a doe that was across the fence that had come out to feed in the open marsh along the creek. She stayed around for a long time and kept my spirits up for sure. There was some rut activity going on this kept me thinking maybe a nice buck would come along to join her. We were allowed to take three deer two had to be 7 points or less and one would have to be 8 or more. This area held some true trophy bucks. I had seen several pope and young deer a couple not at all far from where

I was. Now add in the fact I may also see hogs or even see my yellow pig again ya I was pumped. New stand locations do that to a guy as I'm sure all hunters know. Once the doe fed off not much else happened until an hour before dark. I heard behind me some grunting headed my way. It turned out to be a group of small hogs a mixed bunch of boars and sows some not more then thirty pounds. The largest in the bunch might have gone fifty pounds, all of them I had no interest in taking. They stayed around almost to dark eating up lots of corn then left for the marsh. Once they had gone out of sight I got down and made my way back to the truck. In all the afternoon had been good and left me with plenty reason to be right back there the next morning. Dinner and sleep would have to come first.

Our camp was full that night all of the club members were in along with a few guests. There had been a couple hogs taken that afternoon and one buck. Almost everyone had some kind of activity around them so it was interesting to listen to what they all had to tell. I remember one hunter had seen a monster buck along a marsh and grass pond he hunted next to. He told of the buck running frantic and grunting the whole time he was in sight. His story indicated it might be a buck with a hundred and thirty inch's of antlers and ten points. He had been so enthused seeing this deer he got down at dark and right away moved his stand closer to where the deer had been. This fella was a good deer hunter I was thinking he would be the one to take the first really nice buck of the year. Each year we would as a club take two to three bucks that would go one twenty five or more he might just start things off with a bang taking this buck. Another great time in camp that night would set the stage with enthusiasm for the next day. I was beat and off to bed glad to be in the woods and glad to be a hunter.

The morning found me drinking my coffee talking with my hunting friends about many things. Most we talked about our prospects for our mornings hunt. My drive was a long one so was off and running before most. Even leaving when I did I was running a little late.

Being late left me with a decision to make. I was sure over night all my corn had been found and was gone.

The question was do I take a chance and go now to put out corn or let it go until later. If I was to drive out on the fence line and the hogs were to be just finishing it up I might blow the whole morning. If I didn't go put it out would I be able to stop any of the hogs that might cross the fence later. Well I figured what the heck I was putting it out. This worked out well the corn was all gone best I could tell and saw no hogs in the head lights. Not long I had it done and was on the way to the stand. The sun was just starting to show so I was there in plenty of time.

One of the first things I saw once I could see well was what looked like the same doe from the night before feeding in the marsh. She fed around only for a short while and this time came to my side of the fence to go bed. Not more than an hour later I saw a very large hog I'm sure a boar working along the fence just north of me. I now wished I had extended that corn line up that way I had stopped just past my stand. He came to within a hundred yards and crossed the fence going away. This was good so far it hadn't been light more than thirty minutes. Not much happened for the next hour then finally from behind I could hear hogs making their way to me. Lots of times you can hear a group coming they do talk lots as they move and feed. They popped out right under my stand and right off I knew them. It was the group of small pigs from the night before. They jumped on that corn like it was their last meal. Like the night before they got their fill and were off again to who knows where. I'll admit it was fun to watch them and it does keep the morning exciting. No sooner did they leave and from right in front of me came a buck. He was a young deer a three point. This was a deer I could take so got up and ready. I was sure once he crossed that fence he would stop on the corn and feed. Well he jumped the fence didn't look left or right and as soon as he had hit the ground went trotting off into the hammock. Just as fast as he showed he was gone. Well now I'm thinking this might be one of those days !

Hoping to have seen my yellow spotted hog by now I was questioning where he was. It was past the time that he had shown up on the fence the day before. Maybe he had been there before sunrise and had his fill or wouldn't come along at all. It was warming up and if he didn't show soon it would be warmer then hogs like and be out traveling and feeding. I would give it another hour mostly because I know when hogs find a new free meal they do tend to come back over and over again. Well almost an hour later the spotted hog did show back up and crossed the fence just like the day before. He hit the corn right away but not in my direction. He would have nothing to feed on in my direction the group of small pigs earlier had eaten most of that. I was considering slipping out of my stand and making a stalk on him. If I could get away with lowering the bow and climbing down I might have a chance. As soon as he was facing away I went for it. First the bow then I started down the tree. About half way down I stopped to look and see what he was doing. His head was up looking off in the direction of the hammock to my right. Before I could take another step down he was moving. Right in the direction he was facing off he went heading into the hammock. He went in almost exactly where the little buck deer had gone. Quickly I finished getting down and moved his way. My guess was he was moving right along quickly I couldn't find him at all. My second encounter with the spotted hog was like the first all in his favor. Even having seen him twice now I still wasn't sure how big he was so headed back to where he had been feeding on the corn. That's when I discovered this hog was different. Not only did he have awesome coloration but he had a solid hoof. This hog was a conquistador! Wow I couldn't believe what I was seeing this made me want this hog even more. Somehow I had to make it so.

Back at camp and thinking I had one more day to hunt how would I use that time best. My first feelings were that I had been on that stand twice and in the area a total of three times this might be to much. The decision was to go put out corn that afternoon and hunt a different stand that evening. One of my stands was still to be hunted it had been left alone so far. There was tons of sign around it so the time had come to be there. It also seemed important to be out extra early the next morning and make sure I had plenty of corn on the fence line for insurance. This was my plan and a good one it seemed.

That afternoon I followed the game plan and sat the one stand for the first time. When I walked in the game trails were torn up with sign. That got me worked up right off. This turned out to be good evening I saw deer and turkey and I think some hogs did pass through the thick stuff behind the stand. All in all not a bad day I had seen a lot of critters for sure. On the ride back to came I was already thinking of the morning to come.

Most of us that have hunted for a long time know and appreciate the day to day challenges we face as hunters. We try to do our very best to reduce as many of these as we can. Well on my third morning I created my challenge, I got up late. I hate when that happens. This also is when I hate wanting a cup of coffee so bad. You know I'm fixing my coffee! I hurried started the pot and got camoed up while it perked. When the coffee was done I was ready drank the first cup poured a second and was on my way. With the time being what it was I would be lucky to get there before sunrise much less put out corn. Yes I was mad very very mad. Rather then drive like mad to get there I slowed down and would be parked in time to begin still hunting the corn line. Well if there was any left to still hunt on.

There was a light fog from the cooler temperatures over night. This would aid in me slipping along. When I came to the two track that followed the fence line I could see some of the corn was still there. Where I was standing the fog had lifted. Right away I moved off the fence and out of view. I wanted to be hidden so I could glass and see what if anything was out feeding. Visibility was poor right on the ground out at distance caused from the patchy fog. I would wait it out for a short time to see if anything moved in and out of the fog. Over the next half hour I saw nothing at all. Time to move and if I still saw nothing by the time I made it to my stand I would just get in it and stay put.

My first move would be about fifty yards to the opening in the hammock. This is where again I would have to glass before going across that opening. For the first few minutes there not a thing was in sight. Then out of one patch of fog and into the next I saw the shape of a hog moving. Now this could be good what will it do next. Based on the fact it didn't come out of the fog on my side it either left going away or it was feeding on the corn. Ok I had to make another move I had to know where this hog had gone. Just as I had made about three steps I saw the shape again. This is not good I could be so busted here. It looked to me the hog was feeding so I backed up into the cover again. This hog appeared to be none the wiser I was there and kept feeding. This was doable but I had to find a way past this opening. Looked like the best I could do was back up use the cover to get to the back side of the hammock and cross the open there. The grass in the back of the opening was somewhat taller there and should hide my movement. I did this and made it to the other side but now had to get sight of the hog again from there. There was plenty of cover around me but it was small tree islands, moving here would have to be calculated and slow.

Slipping into one of these tree islands I glassed for the hog. This time I got a very good look at him it was my spotted yellow conquistador. He was right were I expected it to be I found him to be still feeding. He was moving closer to my tree stand with every step. That meant he would be moving into the cover that split the two track in front of it. I could work with this there was a line of trees that were much thicker to use to get to the stand from behind it. The first part of this next move would have to be slow and only when he had his head down to eat I was now only fifty yards from him. I managed to get to the thicker trees stopped and glassed again. He and I were on a collision course. This time as he moved behind some thick myrtles I started my next move to the last tree island I had for cover. This would put me in position for a twenty yard broadside shot. I was almost to that last island when I noticed movement. Stopping dead in my tracks I looked close and from behind some cover I could see it was my hog. He had left the corn line and was now heading right in to the timber my way. I was wide open and stuck with no way to move, this would have to be where my shot would come from. The angle he was walking would give me a shot broadside at fifteen yards or less. He got to that point I drew my bow took aim and let the arrow fly. Well my shot was high it did hit him but very high in the back and right through the fatty area just under the spin. When it comes to wild hogs all this shot will do is make them mad. He had stopped now looking around not sure of what had just happened. Now my nerves were getting to me big time. Pulling another arrow quickly I took my second shot. Well that one wasn't much better I pushed it to the right and straight through his hind quarter. Man what was going on here I'm a much better shot then this and it sure wasn't my first hog. My nerves were shot and this had to have made the difference in my shooting. I was pulling another arrow when I saw this tough old boar reach back and with his teeth pull out the second arrow. Seriously he pulled it out and did this like it was nothing. He turned my way looked right at me he knew exactly where I was now and what I was. He began snapping his jaws like they do when ready to fight. He looked at me as if to say "well now you ruined my day I'm going to ruin yours" next with a really mean look in his eyes he woofed at me and started my way. By now I had my third arrow knocked I drew back when I did he turned to his left and this time finally I made the shot I should have first off. The shot went right through both lungs he went no more then ten yards and fell over dead. Man all I could do was stand there it almost felt as if I had just ran five miles straight, I was spent. What a rush that's all I can say.

Now better composed I walked the short ten yards to my hog. He was an awesome looking hog at that he had the characteristic very long snout and of course the solid hoof of the Conquistador or Spanish hog. He wasn't the biggest in the woods only about a hundred and thirty pounds. I was proud of this guy and the chance to just be out hunting. Any time a hunter can set out after a single individual animal make a plan and see it through to success he can be proud of the effort. This pertains to all critters no mater what. It's non important whether it might be that old mature buck, big four year old gobbler or even the young persons first squirrel.

That was a great year on the lease as most were. Many great experiences for myself as there also were for the hunters in the club. One of the biggest highlights that will always stick in my mind will be my quest for THE CONQUISTADOR.

HUNTING HIGHLANDS LAKE

Our second year of hunting our archery only lease in south central Florida was starting out much different then any in the past. Most of us in this club had hunted this land for many years. There was almost always standing water but most of the time we had ways around it. This year not so much. Land that never got flooded would soon be under two to three feet of water and afternoon rains kept coming.

It was easy as the water rose the critters moved to the highlands and went about business as usual. We would adapt moving stands and feeders inland. But this year even the high country was very quickly becoming a lake. When I say a lake trust me, when you run a Toyota 4x4 with 32 inch tires and the water level is at your headlights on the inland high ground roads it's a lake.

As always we are given the chance to come the weekend before the archery season starts to place stands, set up camps and if we care to pay a small fee we can hunt wild hogs as well. This I always do I love to hunt hogs and every chance I get I'm after it.

One of my best stand locations is right on the edge of a marsh that spans a five mile by five mile area. There are some big bad hogs come out of that place let me tell ya. I knew from weather reports that the rain fall had been excessive in glades county this year and we would be in for some high water. But I never expected to find it this high so early on. We got to camp on Friday around noon set up camp had a bite to eat and loaded up to go put in a stand or two. While doing this the first night out I'll also string out some corn so when I'm done with scouting or placing stands I can go back and hunt for a little bit before dark. My first stop was the hammock along the huge marsh and right away I knew it wouldn't be as good as I was used to. The water line was up to the fence and this I had never seen. The hog and deer trails were there and good sign but how long would this last now long would the water be rising. I had to ask myself these questions before setting a feeder and stand where in a mater of day I may not be able to hunt or even access it. For some reason I couldn't imagine that the water would get worse so off I went and got the area set up. I placed an automatic feeder set to go off twice a day and a ladder stand. Trimmed out some lanes and got out of Dodge. Next I would be looking for higher ground just in case things did get worse. Knowing of a place I hadn't hunted but always wanted to in a high grassy area that was surrounded by oak and palm hammock seemed to make the most sense. There was an old access road to that general area I wanted to look at. This would make it even easier to make the choice. While on the way there I strung my corn. I had spread corn in this area many many times so knew well where and how much to put out in order to maximize my chances.

Driving around the small grass pond leading to the road I was wanting to use to access my new spot didn't show much signs of higher water. This was good maybe I had an idea that was looking better. Finding the old road was easy by the wide hog trails running down both sides of the two track. Well now I'm thinking, this is it and I also strung corn along it. I mean what the hay I was putting a feeder in this spot to and it could pay off right away for this weekends hunt. The hog trails were many and deep with multiple tracks going and coming from all directions. I had still hunted this spot several times and never had I seen the sign or trails like I was on this day. This might just have been an indication of there instincts telling them they needed higher ground knowing the water would still be rising. I found a spot where palm trees had been harvested at least a year before it was perfect. A small clearing bordered on two sides with heavy thick hammock and three small dot island hammocks plus scattered oaks and grass land. This to me was as good as it gets and looked to have all I needed not only the hog trails but deer sign galore. I might just have found my new best spot and I was pumped.

Setting up took me almost two hours, for some reason I just couldn't get it right. The stand didn't give me shooting or the feeder didn't fit where I could shoot and so on. Finally I got it done with still some time to try and hunt.

The afternoon was somewhat warm not much wind and not looking like rain to help cool it off. Still I was off to hunt I spent the rest of the day till dark working my strung corn. The heat never died down the air was almost stagnate the hunt that day was non productive.

Back in camp that evening I talked with several of our members to see what they had found new and exciting. Most had just come to set up camp had their family's with them and hadn't spent much time in the woods. The info I was hearing was water was high and nothing was moving. If it weren't to cool off much that night my corn might still be there. With cool air moving in the morning the hogs might be on it. With this in mind I hit the hay confident I had found a new best spot that in the long run would pay off.

When I woke up in the morning it was easy to tell it was considerably cooler. I felt good about heading out to hunt. Driving into my area I came across a couple small hogs feeding along the corn line they were very small but still this encouraged me. If they were out feeding maybe more would be somewhere along the line. I got parked in my normal spot and off I went. Working the line slowly

I was coming across more deer than normal and not one hog right off. I really hate spooking deer on a corn line it pushes everything. Never the less I pushed on trying to be more careful in my movements, after all it had been all summer since I last hunted. Normally I would have worked the fence line but something was drawing me to my new spot. When I hit the access road right away I noticed the corn was gone not a drop left. Plenty of new sign both large track and small. This could be a family group of mixed sizes. Really didn't matter I was on hogs or at least in the right area. We all know moving into a new area still hunting we sometimes don't watch as well as if we would if we knew it better. This was my problem this morning and I knew this by walking up on the group of hogs feeding on my corn. I got to thirty yards of them and never saw a thing, until they spooked. I stopped to let it cool down after that goof caught my breath looked around more and slowly moved on. Well soon I would be batting a thousand for when I came around the next bend and didn't realize I was right on top of my feeder it was made apparent to me I should have been paying more attention. I heard a woofing sound that wild hogs make when spooked, looked to my right and saw four very good sized hogs running from under the feeder. For ratings sake I cant repeat here what I said but you get the idea. Man oh man I had spooked deer stalking, hogs stalking and hogs again stalking. Must have needed more coffee this wasn't normal for me. You know what the next idea in my brain was, is it going to be that kind of day? I kinda just tucked my tail and headed back to my truck.

This wasn't starting out to good so what now? I figured I should go string some more corn and then check my first stand I had put in the day before. Getting the corn out was simple and took no time at all. I slipped into the stand on the marsh edge and had a look around. The first thing I noticed was the feeder hadn't gone off and the extra corn I had left on the ground below hadn't been touched. Well its getting better all the time isn't it. Turns out the battery lead was off and all it took was a quick fix and back running. Thinking I had made enough noise and commotion I had best head back out. By now its mid morning the heat was up and I was done. Hungry, tired and dejected at my lousy attempt stalking that morning was on my mind so camp was looking real good for a rest and regroup. Things had to be better that afternoon.

Knowing most of the members had gone out that morning I was ready to talk with them to see what all was seen and going on overall. The reports were mixed some seeing hogs many seeing deer but most of all was talk about the high water. Seemed every one of us had a spot we couldn't hunt or in some cases even get to because of it. This could turn out to be one of our most challenging years ever for sure.

Since I was camp leader that year I had a huge aerial map to display. One so everyone could have a good look at the overview but also we could mark stand locations. Many of us had hunted together for some years on this lease and knew what spots most would be in. This year was different we now had the water to contend with and many had to move to adjust to it. To keep the piece it was the right way to go. It also gave everyone an idea what spots weren't being used and might offer a good hunting location. After all this was a seven thousand acre piece of ground with twelve members. Lots of room for all and surly plenty of game once we had it figured out.

Once I had the map up and had a short meeting with the club I had to rest because come hell or high water I was going to be on hogs that afternoon. No way was I haven another hunt go the way it had that morning.

Resting was good the south Florida heat can take lots out of a guy. They were calling for rain showers mid afternoon temperatures would go down slightly hunting could be good. With half the hunt still to go I didn't get dejected I knew this land and knew there's always something around the bend.

The rains came and harder then normal for our mid day showers it turned out to be a full blown thunder storm. The rain had started while on my drive to my area so I decided to park close to my second stand and hunt it that evening instead of stalking. With the rain lasting longer then normal the amount of time it would take to get to and stalk my spread corn seemed better spent where I had seen hogs and had seen them already at the feeder.

My choice this weekend was to bow hunt. When the rain stopped I collected my gear and set out for the stand. Coolness in the air had me pumped up thinking it was just right for an evenings hunt. Slipping in wasn't bad the ground wet and quite. This time I took more time watching to keep from repeating my mornings mishaps. Once

close to the stand I stopped and watched for a bit. I have walked in on feeders in the past while the hogs also were slipping in. This can happen very often right after the rain when the temps have been high and the hogs overheated. Like I said before I wasn't letting the mornings events happen again not now not this time. Not much corn was on the ground from the feeder that morning so I was sure they would come back to feed. Climbed up in my stand I then put up my stuff hung my bow and settled in to wait it out till dark.

I could notice around my stand puddles of standing water it had rained hard so this was normal and may also help me hear oncoming game. Wild hogs don't usually come quite deer do, so standing water helps. Hogs will slosh right through it deer will make light noise but still can hardlybe heard in it walking. This all helps when on stand with limited visibility. Most often in this type of Florida country that's what ya got.

Still kinda tired from the morning I was day dreaming when I heard light foot steps in the water behind me. Surly it was a deer didn't sound loud enough, fast enough or heavy enough to be a hog. I couldn't see a damn thing behind me so until it was right on me I wasn't sure. Then they showed and yes a doe deer and a fawn. Oh well not what I was hunting for but neat none the less. Hey ya never know maybe it might work like turkey season and be a live critter decoy! Straight to the feeder they went so obviously they had found it the night before. This was good the feeder had been discovered by both hogs and deer, things were looking up.

My friendly deer stayed around for only about half an hour as most often they do. Plenty of time was left and the feeder was about to go off any time. In the hog woods when the feeders go off we call it ringing the dinner bell. Their loud when spinning corn out and can be heard for a long way off. I have seen wild hogs come on a dead run out of a dead sleep when it happens. This is one of the times you will also hear them very vocal on the way in.

That dang feeder went off and about scared me out of my wits! I knew what time it was going to I was ready for it to and still it got me. Hate when that happens. Now the real anticipation comes you know it went off had to be heard a long ways away so be ready. I waited and waited and waited nothing happened. Waited a little longer still nothing. Learning is always a part of every hunt and I had learned many years before to use feeders that come with settable timers instead of automatic types. The auto types go off a half hour before dark and if your hogs don't come right in then your shot may be coming almost to late. I recommend and do set mine to go off from an hour to hour and half before dark giving plenty of time for them to show. I really hate tracking wild hogs in the dark and in the woolly nasty swamps or marshes in south Florida so it's maybe a self defense thing.

The sun was below the trees when I heard what I was sure were several pigs heading at me. Again in the standing water it was apparent what they were. To my right was the small clearing and they were coming across it and coming fast. Four in total It had to be the same ones I had spooked that morning. All were in the hundred and thirty pound range looked to be boars very fat and healthy. My bow was in hand I was standing and ready.

This is when it gets intense. I knew the yardage they would come by dead broadside at twelve yards. If it had been earlier I would almost always let them dive into the corn as a distraction to my movements. This evening not a chance time was short light was getting dim so when the chance would come I was taken it. Three took a slight turn to their right, away from me slightly the other one still came straight in. The ones that turned were coming to the feeder from a different angle almost quartering towards me but still the other was dead on coming broadside. I drew back the bow, took aim let the arrow go and watched as it hit home. The pig took off running hard out fifty yards made a full circle came right back at me and fell over right in front of the stand. Wow no tracking this time, works for me !

Some days everything goes right some everything goes wrong this day was a mix of that. Lessons were learned time in the woods was awesome and at the end a nice wild hog to show for the efforts. While loading up my pig I kept thinking I have a full day left to hunt maybe just maybe I'll get another chance.

Dinner that night seemed extra good and sleep would be welcomed. I had finished the work up on the hog before dinner. Now time to rest and maybe dreams of a sit on that stand again would come as well. Not a bad day, not a bad ending and not a bad way to feel.

Morning came way to early but I got up. Making coffee was first thing to do, no coffee no hunt. With coffee in hand I sat for a bit and enjoyed the cool early morning air. It felt good felt like a morning that might have the pigs moving. Not sure how it would go at least I knew it wouldn't be to hot to start with. Got dressed loaded up and headed out thinking if I were to be lucky this morning I could head back home that afternoon. My plan had been to stay until Monday noon but if I were to have two hogs down I would have no need to. My stands were set feeders in and working one getting used already so I was set for the opening weekend of archery. If the second hog were to come on this day that would just be the icing on the cake.

My biggest problem driving in was where I wanted to hunt. Should I go to the first stand, should I spot and stalk or should I go back to the second stand I had hunted the night before. These problems maybe everyone should have on a weekend morning in the hog woods. Even when I got parked I had no clue where I was going. I really didn't want to spook anything off the most active stand plus I had taken a pig there the night before. I hadn't seen much corn left on the ground I had strung out so that wasn't looking good. The first stand along the marsh was looking to be the best bet. Something by now had to be on it just wasn't sure if so when. That would be my choice right or wrong, it would have to do. Gear ready, enough light to move without a flashlight off I went.

Slowly I got to the fence that I had to cross to access the swamp stand. I had strung corn the first night along it and was sure it would be gone. Sneaking in I saw several smaller pigs working along heading away from me. It didn't look to me as if they were feeding just heading back to cover. This did make me think I had to be extra careful walking into the feeder and stand. One never knows and it's very dense most of the way in, care would be needed.

Since I knew this area well I made the decision to loop around from the left side instead of from the right. This would give me more cover to hide myself and possibly let me get a look under the feeder from a distance. The plan was a good one and worked well but to no avail. I got to where I could see under the stand and nothing was there. Even though I wasn't far from it I couldn't see if any corn was left so I would still need to have a look. Maybe even re-set off the feeder in hopes of attracting back a hog or two. I have done this many times before and sometimes it works. The feeder had been hit and it looked by track size by some very large hogs. I got pumped maybe they were still close and still hungry. Quickly I did just that set it off again hurried up the stand and got ready. Now the waiting game is on what will come, when will it come.

Wasn't long and I could hear something coming. Trotting sounds were coming from the marsh and sounded like it could be the large hogs that had left all the big track sign. Had my plan worked, well sure seemed like it had. Now if they continued on the path it seemed they were on it would be hard to see them until within twenty yards of the feeder. This was ok at least they couldn't see me as well. I'm sure like most hunters I was caught up in the moment and really hadn't noticed the wind. This would soon be my downfall.

The first hog showed stopping right on the edge of the thick cover. It was facing almost straight at me and checking out the open area. This didn't look right something had him spooky. For them to have come all that way in a hurry and to just stop got me thinking whats up. Probably just as fast as they appeared the first one woofed, spooked and off they ran as if shot from a gun. Just siting back down was about all I could do. Wasn't much I could have done different this time. I dealt with what I had and went with it. At least now I knew I had activity on both stands this should make the following weekend even better with options to hunt. Staying for a little while longer gave me a chance to look about from the stand and see if I needed to do anything more. I had not cleared out any shooting lanes much and decided that I wouldn't need to. Everything would have to go in front of me and it was wide open. This would work just fine.

Back to the fence line I went it was still early enough to catch something moving but I really didn't think it would happen. There was a huge live oak that stood right on the edge of the two track that followed the fence. It would be where I could stop to have a look down the fence. When I first got to it there was nothing. In no big hurry I stuck around just watching and thinking it hadn't been to bad a weekend I might just head back home thankful I had been in the woods. When out of nowhere came three big hogs. I didn't know for sure but felt it was the same ones that had spooked from me not long before on stand. They had come out from the marsh side and in the general direction of where the ones had run when they spooked. Never the less they were coming my way and fast moving.

Hidden by this large oak would work if nothing changed. Steady they came slowing some and on my side of the fence. I was ready for the shot just a little closer now. In single file walking eight yards away broadside was perfect I drew back let the arrow fly the shot was true and my hog ran dead straight down the fence until dropping some seventy five yards away.

The other two ran back in the direction of the marsh and seemed not to really know what had just happened. Wild hogs are smart to begin with, for them not to know whats up is best. They do learn very fast what hunting pressure is.

This had turned out to be a great weekend of hunting and getting ready for the season to come. I'm always ready to go to the woods no mater what. To also have success while doing so is a plus for me but not the important thing. In our fast paced lives these days often we can lose sight of the simple things that make life good. Maybe someday your hunting success will be in a place like I found while HUNTING HIGHLANDS LAKE.

BIG BOAR BY PISTOL

For many years I had hunted wild hogs with a bow mostly. I had taken a few with a gun but bow hunting was my passion so it was almost always my weapon of choice. Also up to that point I had always done my hunting on public land. We have been allowed to hunt hogs year round in Florida forever but unless you had a privet lease you would be stuck to only during the hunting seasons. I had always wanted to go on a really good hog hunt on privet land for a really big boar. My chance was about to come.

Late in the gun season this particular year I was talking to one of my friends about how the hunting was going on his lease. The lease members hadn't taken many deer but were knocking the hogs down like crazy. I told him I hadn't yet that year harvested a single pig. Hinting that I would love to take a good hog for the freezer my buddy told me that maybe soon I could come and do that.

It was getting close to the holidays when I received a call from my friend. We had talked for a short time when told me that he had talked over with the lease members about me coming to hunt. Most of the members were about done hunting before the holiday season that meant the lease would be wide open for me to come hunt on a weekend. We made our plan picked the weekend we would hunt, I was pumped.

Friday night before the hunt we again talked on the phone so I could get directions to the property. I hadn't spent much time in the area his lease was at so had to make sure of all the details. We also talked some about the size of the property and how it laid out for hunting and what the hogs had been doing. You know the normal around the campfire stuff just it was on the phone. I told him I would be bringing my 270 to hunt with it would handle anything that might come along. From what he told me most of the shooting would be fifty yards or less at hog feeding at feeders. This was perfect I was looking forward to this new experience.

My friend met me at the gate well before daylight. He wanted to fix some breakfast and coffee before we headed to the stands. This sounded great to me especially the coffee part. I knew his cooking and knew it would be a full spread this would be good fuel for starting the day. While he cooked we talked about hunting that year some of how it had been for me and some about his plus how it was on the lease. From what I could tell this place was full of hogs and maybe even to full. The land size was less then fifteen hundred acres they were running twenty five feeders and every one of them were getting hit every day twice. This was impressive to say the least. As we ate our breakfast he asked if I had brought my 270 I told him I had. This is when he made mention of what he had not mentioned before now. This was a hand gun only lease and my 270 would be staying in camp.

I knew that the members on his lease did hunt lots with hand gun but had no idea that it was hand gun only. My friend just kind of smiled at me and went on to tell me he had just what I needed. He would be loaning me his single shot Thompson Contender chambered in 35 Remington 14 inch barrel topped with a 1.5 X 4 power scope. You can bet on two things one I was a little apprehensive and two excited to try this new way to hunt. He told me the gun was dead on out to a hundred yards I should have no problem hitting what I aimed at. Well I guessed we would find out soon enough wouldn't we.

After going over the pistol with me we talked about the stand I would be hunting. There were several good hogs coming to the feeder at this stand and that the shooting would be at fifteen yards. This sounded to me to be just right for a first time try with a handgun. We loaded up and headed out to our stands. When we arrived where I would be on stand he walked me in to find it. The false dawn was just starting and the feeder was set to go off an hour after sunrise. Plenty of time to get orientated to my surroundings. My thoughts while waiting for the sun to rise went from apprehension to wow this might be a blast. I was ready.

Not long after it became light enough to see, what came by first were two doe's. Doe's weren't allowed to be taken so best I could do was watch them as they moved through. This was still a good start with game moving by

right off. With still another half hour to wait for the feeder to go off I was content to sit and take in the morning. This stand was set up in a very thick sweet gum tree hammock. The ground was covered with lush vegetation that helped define the wide hog trails leading to the feeder. This was helpful for letting me know where they would possibly come from. All around for ten yards under the feeder there was nothing but dirt and standing water. Again this was good it would allow for wide open shooting. With not much time left to wait on the feeder my comfort level was much higher knowing more about my area.

No mater how much you anticipate a feeder going off it always gets ya. This one got me for sure I jumped about out of the stand. I settled back in knowing the hogs would be along very soon. If there was ever a time I had to be most on top of my game it would be now using a handgun I had never shot. Handgun hunting for hogs that also I had never done. While your sitting there in wait many things go through your head. How bad will this thing kick or will I be able to hit my spot you might even be thinking will this handgun do the job. I did know a lot about the 35 Remington I had that same round in a leaver action riffle. It was one heck of a hog gun so I felt sure it had to be just as good at close range out of this

Thompson. Would I hit my target well I had shot plenty of handguns so that I felt would be no problem. As far as the kick all I had to go on was how it did in a riffle the 35 does kick. I'm thinking this must be why there is a fore grip on this thing to help keep it down and under control on recoil. Other then all that no worries.

It wasn't long and I could hear something coming right at me. Just as I first heard them I got sight of 4 good sized hogs heading right to the feeder. It was clear they were on a mission to get to that corn. With no caution at all they came straight in. Everyone of them started on the corn on the opposite side of the feeder from me. This gave me no shot except head shots. I was not comfortable in taking that type shot with not ever shooting this Contender. I would have to wait for something better for an angle. These hog fed for no more than eight to ten minutes that's all it took to clean up all the corn. Never did they do anything but stand facing me leaving me with no choice for a shot. Without warning they turned around quickly and trotted back from where they had come form. Just that fast it seemed to be over. Wow that I didn't expect yet still it was ok I felt good about not taking a shot I was uncomfortable with. The morning was still young there might yet be another chance.

Probably not more than ten minutes later I heard a

shot from the direction my friend said he would be from me. On the ride in that morning he told me he was going after a smaller hog for the grill. He knew exactly the feeder and the group of smaller hogs he was after. This sounded good to me that would make for some good eating that night after hunting. Thinking of good eats later I went back to surveying my area. My hunt this morning wasn't done yet there was still an hour before we would meet up out on the road.

The next hour was spent watching and waiting with nothing going on. Close to time to get down I could hear my buddy coming from behind he was walking in to get me to help him drag out his hog. I was good with that the corn was all gone and it surly was later then the hogs would like to move. On the walk out I told him of the mornings events. He assured me that all I needed to be was patient the shot would come. Off to his hog we went, on the way he told me the story of his hunt. Just as he had expected the group of smaller hogs did come in and he took just the right one for our cook out that evening. The stand he had taken the hog at was also explained to me to be a good evening stand. From what he told me the stand I was on that morning didn't do much in the evenings. He had wanted me to see it just in case I wanted to make a move for later that afternoon. After working up his hog and lunch we would drive around and look at other stand locations and make the final decision on where to hunt later. I liked that idea welcoming the chance to look over more of his lease.

Much of this land was more open ground with pines being the primary timber. That part of the lease was better suited to deer then wild hogs. We took the time to go over the entire piece of property and looked at every stand and feeder. It looked like the best places to be were stands closest to the heavier or swampier areas. This is where the most and freshest sign was found and always had been where the most hogs were harvested.

The last of our stops would be along the back fence line. This is were my friend thought I might want to be for my evening hunt. There were two stands along the road that bordered that fence. Both were getting torn up buy the hogs. When we got to that fence my friend *told me this is where you might just see anything and winked as he said it. Ok I had to ask what did he mean anything. That's when he went on to explain just what exactly was on the other side of that fence. Well turns out on the other side was an exotic game ranch. That place was loaded with everything you can think of from full curl ram sheep, bison, huge trophy wild hogs to trophy bull elk. He then went on to explain to me that any of these that were to come on this land would be considered fair game. Their considered non native game by the state and were legal to take year round. My friend had just two days before had an elk come down along that fence but never had the shot. Ok that's it a no brainier I would be sitting right there that evening. Oh the hog sign was intense there as well. Back to camp it was for a short nap our plan was set.*

After resting a short time we headed out to the target to take a few shots. It was best I did to know just what I could do if the chance were to come about. I'll tell ya it was a blast shooting that Contender for the first time. I was hooked from that time on about single shot pistols. Once I was comfortable shooting it was out to the stand I had picked. The thought of a bull elk or ram sheep and maybe even a true trophy hog coming across that fence to a feeder had me mesmerized to say the least. My friend did tell me this didn't happen much but you never know it might at any time. He dropped my off at the stand I got up in it and ready for come what may.

Right away the first thing that I noticed was that my shot to the feeder was close to fifty yards. Most of the hog sign came from behind me and along the road almost right under me. That would be good if they were to come that way it gave me shooting at ten yards or less. Everything was looking good to me now the waiting game would begin.

We had put out a line of lose corn just in case something were to come along before it was time for the feeder to go off. This could also slow the hogs down on the way to the feeder and that possibly giving me a closer shot if needed. The next couple of hours were spent watching and waiting. About the time everything starts to come alive I got one heck of a surprise. Across the fence came the sound of a bull elk. He was some distance away what I figured to be three or four hundred yards. He was bugling something I had only heard on TV or on a video and never had heard in the wild before. This was awesome I had to hear it a couple more times before it really sunk in how majestic that sound was. The rest of that afternoon was spent sitting and waiting on the hogs that never showed. It really didn't mater the sounds of that elk made the slow evening well worth the time. Soon my ride would show and it would be off to the next good thing Dinner.

My friend had not hunted that evening and stayed in camp to cook the hog from that morning. This was going to be one heck of a meal with all the fixings. He made fried potatoes, baked beans, pan biscuits and some kind of dutch oven desert. Yes we ate way to much and had to sit around the campfire to recover. This time was also spent talking of times in the field gone by. We also made the plan for the next morning. I would be going back to the stand I had hunted first. After many stories and talk of the next days hunt it was off to bed.

The morning would find me still full from the night before. I opted for just coffee I had no room for breakfast. The morning air was very cool and the sky's clear this had the makings of a good day to hunt. We had our coffee grabbed our gear and were off to the stands. This time I would go to mine on my own my friend dropped me off and went his own way. The trail in was easy to follow so I made it to my stand in no time shortly was up in it and ready. While sitting in the dark waiting for the sunrise I had that feeling you get when something was going to happen. I'm sure this made me even more aware of my surroundings. As regular as it was that these hogs came to the feeders I was sure the four from the morning before would be back. This time I would be ready for any shot I would get.

The false dawn came and went next it would be time for the feeder to go off. When it did it got me again yes I jumped. Right away I was watching the trail the four had come in on before. I waited and nothing came. I stayed ready I knew they would show any minute. Then out of the corner of my eye I saw a big black shape coming. This was a big black hog it was on a different trail coming along very slow very cautious. That hog was not one of the four I had seen the day before he was much larger and way to nerves. This hog knew what he was doing and was in no rush to get to the corn. He had to be an older boar one that was well aware that the corn was there because of man. If the opportunity were to arise this would be one heck of a wild hog to take. When he got close enough to see his head I could see his cutters and they matched his size. When he got within fifty yards of the feeder he moved side ways paralleling it and me. I could tell he was making very sure it was safe all around the feeder before committing to the feed. I had good wind on him I just had to wait him out. Finally he stopped out at fifty yards partially blocked by a tree. This is where he stood for almost ten minutes. He never moved a muscle and hardly even moved his head. He was well within range but I had no shot. There was no hurry I felt he would come in on his own time.

Very slowly he made his move in my direction. I was as ready as I could get. With this hog being so big my shot placement had to be right. He would have to give me what I needed to make it so. For this guy to come to the feeder he would have to turn slightly to my left this would still not give me the angle I needed. It would have to come at the feeder if he were to move around it. First thing he did when he did finally get to the feeder was look all around again. Once he was satisfied things were ok then he went to feeding. At first he was facing me dead on like the four the day before. This was not good I needed him to turn. Slowly he started to my right but still quartering at me. Then suddenly he turned hard left that was just what I wanted. I got the contender up and settled the cross-hairs on him and took the shot. His feet went right out from under him and he hit the ground like a ton of bricks. I was stunned to say the least. Just in case I reloaded and kept a very close eye on this guy. For the next couple of minutes I just couldn't believe the impact the 35 had on this big old boar. I mean he just collapsed and never made another move. Just as I was about to relax he jumped up turned 180 degrees and took off making it about ten feet and ran head first into a tree. Down he went again. Now I was sure he was done and that was his last effort. Never the less I this time kept the gun on him and ready this time for a head shot. Ill bet it was five minutes I waited and nothing happened. Then it happened again he raised his head when he did I took the second shot. His head hit the ground hard no way

could he would he get up from that. Within a few second he started kicking surly it was his death kick. The force of his kicking moved him part way around the tree. I still had a full body view if I needed to make another shot. I hadn't reloaded after the second shot but while sitting there I figured I might ought to. This as it turned out was a real good idea because as I shut the breach from reloading he stood up like he had never been hit. Oh my gosh what the heck is this? Not in those exact words but you get the drift. I took quick but careful aim and made my third shot on this boar he bolted and ran right into the same tree as before this knocked him right off his feet, again. Ok I was stunned before but now even more so how the heck did he keep getting up. Well if he ever stays down and I get a chance to have a look I'm sure going to find out how he did it. Reloading for the forth time to be ready seemed to be necessary. I waited for a few minutes and nothing then waited a few more minutes still nothing. This was starting to look like he was done I sure hoped so I was running out of bullets. Sitting it out for another thirty minutes with not a hint of movement out of him I felt sure it was time to get down and have a look.

When I got to the ground first thing I noticed was he was as big as I thought maybe even slightly bigger. Making my way to him very slow and watching him very careful I could tell he was not breathing he was done. Now to find out what the heck had just happened. How on earth did this guy take three rounds to keep him down. Well the first thing I noticed was the hole in his ear. Ya in his ear slightly grazing his brain cap. My guess was the second round was just a grazing shot but still enough to knock him out for a few minutes. Then I found the first and third shot placements. Seems the first round hit him in the brisket but not a fatal hit. This would have been enough to have knocked the wind out of him and knock him to the ground. It might even have been possible it put him out for a few minutes that's why he didn't move for that four or five minutes. The third shot hit home just right it was a perfect heart shot this put him down last and for good. The first and third shots were only two inches apart. So all in all the first shot was almost perfect. The second shot couldn't have been more than half an inch off. The third shot was perfect. For the first time shooting a single shot Contender I'll take it. It wasn't the best I could do but it did get the job done. And this was a heck of a good hog to be the first with a handgun.

It was sure my friend had heard all my shots and soon would be around to find out what had happened. I headed up the trail to the road to wait on him to arrive. Not long after I made it to the road he was there and back we went to retrieve my hog. Once he heard the story he then understood all the shots. He got his chuckles out of it and we headed back to camp.

I would hunt that evening but had no luck. What did happen was I got to meet the vacuum cleaners. Yes the vacuum cleaners I said they were a group of nine small pigs around fourty pounds each that made the damnedest sounds when they came in to

feed. Their name was well deserved for sure. When they got to the corn like a revolving wheel they started made one complete circle sucking up the corn like candy. It sounded just like a vacuum cleaner sucking up rocks and by the time they finished that one circle all the corn was gone not a single kernel remained. Damnedest thing I had ever seen and all I could do was laugh like crazy once I saw it happen.

My first handgun hunt for a wild hogs was truly a blast. I was hooked and eventually bought that exact same Contender set up identical to my friends. This is a sport I do recommend to all. The challenge it presents the close range nature of it like bow hunting and being single shot makes it awesome. I thank my friend for introducing me to it. In one short weekend I was turned on to single shot pistol hunting and took my BIG BOAR BY PISTOL

A LONG SHOT

One of my late spring early summer traditions was always to go on a wild hog hunt in south Florida. The place I would always go was along the famed Fisheating Creek. This hunt camp had four leases that a hunter could book two day hunts on at a very reasonable price.

These hunts were do it yourself type hunts and very good ones at that. The format was to camp on your own arriving on Tuesday at noon leaving on Thursday at noon. You were allowed two hogs of any size. This particular year I booked my hunt on the Creek lease. What made this hunt extra special was I would have the whole place to myself. Over six thousand acres for two days that was flat loaded with big hogs.

I had many times guided both Wild Hog and Turkey

hunters on this property but this would be my first time hunting it for wild hog on my own just for me. This hunt I could hardly wait for I knew how good of an area it was and that it held some monster boars. My overall plan was to take a meat hog and then go for as big and nasty of a boar as I could find. My weapon of choice on this hunt would be my trusty 270 Winchester.

The day of my hunt I was at the camp office at noon did the normal formalities and headed out to the camp. I brought my small camper to stay in so setting up camp was a breeze. With it now being daylight savings time there was plenty of time to get out scout and put some corn down. Once I had this done I could decide what area I wanted to hunt first. From talking to the lease manager I learned that one of the big pastures along the lake levy had lots of hogs coming to it before dark. This was one of the places I put out the corn and would be the first place I would hunt. My plan was to park at one end of the field walk along the levy to the mid way point of the corn and hide in the paper trees until dark. Everything around was very flat so I could see a long ways from that point.

With everything done I got myself camouflaged up and headed to the pasture. I arrived at my parking spot around three thirty that would give me almost five hours to hunt. With so much ground to watch it was sure that time would be full of things going on. This was a game rich place not to many days I had spent there were without regular activity. Once I was where I wanted to be the first thing I saw was a big flock of turkeys. This was a flock of hens with their poults out feeding on the pasture I'm sure for bugs and seeds. It looked to me like the hatch had been a good one judging by the number of poults per hen. Very good information I could take back to the lease holders who were very well known turkey researchers. Next came several deer to also feed in the field. So far everything had come from a very thick swampy area this was sure to be

where the hogs would come from to. Many of the swamps in this area held water year round. This would make them some what cooler during the warmer days plus provided plenty of cover. The area many many years before had been timbered for cypress but the new growth cypress made it very thick and in some places almost imposable to hunt. The swamp across from me was just such a place. The time spent just looking over this almost untouched land can make one wounder of what it had been like at the time of the early settlers. I know from history of the area parts of it had been settled in the early 1800's. It's almost imposable to imagine how tough it had to have been in those days. You almost can feel a small connection to those who came before. What you can never do is feel the true hardship they underwent. Most of my first evening was spent thinking about just such things. Now with just two hours left until dark it was time to concentrate on hogs.

I really wanted a good sized meat hog for the freezer. My hope was that would be what I saw first. It wasn't long and it looked like that wish was going to happen. Straight across from me coming from the swamp were a couple smaller hogs. That couple turned into a few then into many. They were of all sizes from thirty pounds up to what I guessed to be a hundred and fifty. Two or three of them were exactly what I was looking for, the perfect size. This was a great start they were already on corn with still plenty of time I would wait a little longer to see what else might show up. Over the next half hour several more hogs came out from different directions none of them were any better then what I was watching in the first group. Mater of fact most of them were boars and for a meat hog I wanted a sow. My decision was made it was time to pick one from the first group. Out of the three or four big sows one seemed to be without piglets and slightly larger that would be the one I wanted. Placing my back pack in front of me for a rest I got ready to take the shot. This shot would be at least a hundred and fifty yards. I settled the cross-hairs on the hog I wanted but had to wait for a good broadside shot. This didn't take long it moved to the left I took careful aim a deep breath and squeezed the trigger. The shot was true she went right down all the rest ran off I had my first hog of the hunt. Standing there and watching her for a few minutes told me she was going nowhere there was no need to go ground check. Right to my truck I went and headed back to collect her. I was pleasantly pleased when I got to her she was every bit of a hundred and fifty pounds and would be perfect table fair. Very satisfied back to camp I went, half the quest done.

Once I was done working up my hog it would be time for dinner around the campfire. Being alone eating dinner around the campfire in this most wild land keeps the mind wondering full of thoughts about times gone by and times still to come. No way no how can life be any better for a hunter.

My first morning started like normal coffee, coffee and then some coffee. I had put out corn in several locations the previous afternoon. The morning hunt would be along those locations other then to the pasture from the night before. The plan would be to drive to each location park and still hunt into them. From here on out I would be selective I was after the ugliest, biggest, nastiest boar I could find. I had the entire day morning and evening to find that boar. If this didn't happen I would be back to the pasture the next morning my "ace in the hole" for another meat hog the very last thing in the hunt. It was time to head out and put my plan in motion.

Once I left camp I had to cross the creek to get to the spots I wanted to hunt. From the time you got on the other side it was hog heaven from there on. I planed the time I left camp so I would be entering that part of the area just as you could drive and see with out the use of the headlights. Many many times over the years I had been able to do this and see the shape of a hog park and still hunt after it. This is not a bad way to find that older boar before he heads in to nest for the day. The older boars don't stick around much after daylight breaks they know better. Creeping along I saw the first hog it was very large, alone and heading to cover. This had to be an older boar and just maybe the true trophy I was looking for. He didn't look like he had noticed me coming but they don't always let you know so I parked and went after him. Falling in behind on the trail he was using I slowly moved along. At first there was no sight of him all there was to go on was his track. The track told me he was well over two hundred pounds so just might be the trophy I was looking for. We played this cat and mouse game for over a hundred yards. Just as he made a corner heading into a tall grassy area I got my first good look at him then he was gone. Some how I had to make up ground if I were to have a chance at him before he hit the thicket he was almost to. Instead of following the trail I cut off and went right for the grassy patch where I last saw him. Somewhere close to that spot he had cut off the cattle path we had been following. Now there was no track and no sight of him surly he had made it to the thicket and safety. I was disappointed but this was a good start the hogs were out and moving and I hadn't even got to my corn lines.

My next stop would be at to park at a gate along a fence line. The road lead into a pasture and snaked around several small cypress swamps dotted along through it, on the west side was the creek hammock. This should be an area of travel that I had put a lot of corn out all along it. The idea was to still hunt along

and do lots of glassing. Because it was an active well used pasture the grass was very short and not much vegetation around the swamps. Even if the hogs were heading back in to lay up I should see them well before they got to the heavy stuff and out of sight. For an hour I moved slowly along and covered the entire area. Most of the corn was gone but not a hog in sight. I had one more line to cover before the sun would be to high. By the time I would be done with that line it would then be to warm and surly the hogs would be back hiding in the shade.

Spending another hour slipping along the last corn line produced no results except to tell me I would be spreading out more corn later that day. After I went to the pasture where I had taken the hog the night before and putting out some corn there I was off to look over more ground. This I did until ready for lunch I saw some great country but nothing I wanted to hunt more than I already had. It was time to eat.

After lunch and a nap I was off to put out more corn then hunt. First I re corned what I had before and rather then a line I was spreading it like seed. This would take them longer to find it and longer to suck it all up. I hoped this would keep them out in the open longer. When this was done I had a small opening I wanted to corn because of the huge tracks I found on a trail that cut across it. The ground cover was a little taller the cattle had not been feeding there. I figured it was a huge boar crossing it at first light to get to cover using the thicker ground vegetation as cover to get to it's bedding cover. It was a long shot that the corn would last over night but I had to give it a try. I had to carry the corn in by hand it couldn't be driven it being to wet and all. Even taking the time to do this it was still very early. Much to early to just go and sit so I headed back to the start of my corn lines. There have been times I have done this and already the hogs are on the corn. It was worth a shot plus it would waist some time. It took some time to drive all the lines slow and didn't do a thing for me so it was off to the big pasture from the night before.

After driving around the extra time I did I was arriving at the pasture not long before the time that everything had happened the evening before. By the time I would be to my hiding place it shouldn't be long before the hog would be out. Before walking in I took time to glass all the pasture just in case the hogs were already out. So far I could see nothing out. With so many hogs around there you never know what might come out that hadn't been seen the evening before. My hopes were still to see that nasty looking monster boar. Oh well nothing out yet so off to my hide I went.

Not long after getting to my hiding spot two hogs came out from behind me. I had started the corn line just past where I was sitting. They fed along the line right in my direction for most of the way. They left and out in the pasture they went. Both were smaller then the one I had taken the evening before so were of no interest to me. Way across from me I could see several deer coming out and still there was no sign of the big group of hogs I was expecting. The evening stayed just that way and as dark fell I headed back to the truck thinking in the morning the first good sized pig I saw was mine.

While watching the camp fire that night my thoughts were on the small opening I had put the corn out in and hadn't hunted yet. The best plan seemed to me to be on the move like I had that morning but if nothing was seen to keep going right to the new spot. I could get up on the levy in the paper trees and watch over the spot for hogs to be crossing it. If there was any of the corn left then that was a plus as well. This seemed to be the best idea. I had to go past all the corn I had set out on the way to that new spot I would either see hogs on the way or not. If I wasn't seeing them out along the corn most likely it would be a waist of time to hang around. The focus would be on the new spot.

My second morning started by consuming massive amounts of coffee. This would be a short morning I had to be out of camp at noon. The night before I had somewhat put up most of my camp it would take very little time to finish up after the hunt. That would leave plenty of time to hunt and be out of the woods by the time the hogs would be back in cover. Sure I was sufficiently loaded up with coffee I headed to the woods. Ok I like my coffee what can I say !

Following the same route as the morning before I was watching for any sign of a hog I could find. There was nothing to be seen anywhere. This got me to my new spot in plenty of time. I parked and went right into the paper trees that lined the levy. Slipping along the levy was almost like being in a tree stand. I was well elevated overlooking the small spot I had corned. This would be perfect I could see to shoot the entire area. Finding a good vantage point to sit and wait I got ready for whatever might happen.

The spot I was overlooking was about twenty acres

in size and almost square. One end was open the entire width of the opening the other end had a line of myrtles that separated it from an open cattle pasture. I was able to also see out into part of that pasture a good ways. From this hiding spot I could move in any direction and be well concealed. Now all I needed was for a big old hog to show.

Over the next hour not much happened expect for a couple deer passing through. My gut was telling me I needed to go have a better look at the cattle pasture. Seemed to be the thing to do so off I went. When I got to the line of myrtles there was a cattle trail led through to the other side. Slowly I snuck into the pasture using that trail. Once on the edge I could see parts of the pasture I hadn't seen from the levy. From this point I glassed it all. Off to my left was a slightly

hidden corner. As I began to glass over it I saw a pair of ears. It was a set of big black wild hog ears. Glassing further I saw the butt end of another hog. Both were feeding and seemingly in no hurry to go anywhere. Within a couple minutes both were in the open one standing broadside. This hog was about fifty yards out and looked to be at least a hundred and twenty pounds. Another perfect hog for the freezer I was going to take it. Raising my 270 and setting the cross-hairs on the shoulder I took my deep breath and squeezed of the round, my hog hit the ground. Ready for a follow up shot I waited for a couple minutes just in case. The need for that follow up wasn't needed so I was off to have a look. This was a good hog a sow in very good shape and a good size for great table fare. I was pleased plus I was done with my hunt with plenty of time to spare. I could drive within a hundred and fifty yards so the drag to there wouldn't be to bad. This I had done in no time and was off to my camp.

Working up the hog and packing up camp didn't take long at all. I went to the hunt camp headquarters checked out and headed back home. This had been one heck of a hunt the result of it being many great meals to come another great time afield and good stories to tell. What a great experience one that I hope every hunter at some point in time gets to have. Remember to always be prepared you may have to take that hog with A LONG SHOT.

AUTHORS WORDS OF WISDOM

We all as outdoors men and women have an obligation to give back to our sports more than we take. No one should ever take for granted what we have in the wilds we so enjoy today. The true successes in hunting and fishing don't come from the kill its from the sharing of the sport. A short hundred years ago many of the species we enjoy today were all but extinct the Bison and Wild Turkey prime examples. Many millions of dollars and countless man hours later wildlife today is thriving.

Whether your a Turkey hunter or Water-fouler or maybe you hunt Elk, Whitetail Deer or Wild Hogs there is for almost every critter we hunt or fish an organization that represents them. No mater what your interest is join your local chapter or national club. Volunteer some of your time we all have something to offer. Please always take your sport to heart.

Last but not least mentor a new or novice hunter weather its your wife or girlfriend husband or boyfriend maybe your child or the kid next door the benefits for the sport and the wildlife of all kinds can never be measured.

Richard M. Schamber

FOLLOW RICK AT
THESE LOCATIONS

Pro Staff Member
Join me at
http://www.huntskills.com/

Pro Staff Member
Join me at
http://buckmafia.com

Twitter

http://twitter.com/authorricks

Facebook

http://www.facebook.com/profile.php?
id=100000223375830

WEB SITE

http://authorrickschamber.webs.com/

AUTHORS BOOKS AT

http://www.amazon.com/

AUTHORS PRODUCTS

http://www.zazzle.com/authorrickschamber

CHEROKEE DREAMS PROMOTIONS

http://cherokeedreamspromotions.webs.com/

Special Thanks To These Fine
Outfitters

For Information On Hunting
Missouri/Iowa/Texas

Steve & Donna Shoop
J&S Trophy Hunts
(641)-724-9150
http://www.jstrophyhunts.com/Index.html

My Georgia Connection

Nickie Roth
Archery Outfitters
(727) 525-2825
http://www.archeryoutfitters.us/index.cfm

PLEASE SUPPORT
THESE CAUSES

PLEASE
CHECK OUT

MAGNOLIA GAME CALLS
http://www.magnoliacalls.com/

FIELD NOTES